Yale & the Strange Story of Jacko the Ape-boy

The No. 4 train tunnel near Yale, British Columbia, close to where the strange story began on a summer day in 1884.

Yale & the Strange Story of Jacko the Ape-boy

Christopher L. Murphy
in association with
Barry G. Blount

ISBN 978-0-88839-712-6

Second printing 2011

Cataloging in Publication Data

Murphy, Christopher L. (Christopher Leo), 1941–
Yale & the strange story of Jacko the ape-boy /
Christopher L. Murphy in association with Barry G. Blount.

Includes bibliographical references and index.
Issued also in electronic format.
ISBN 978-0-88839-712-6

1. Sasquatch--British Columbia--Yale. I. Blount, Barry G
II. Title. III. Title: Yale and the strange story of Jacko the ape-boy.

QL89.2.S2M88 2011 001.944 C2011-901299-5

Printed in South Korea — PACOM

We acknowledge the financial support of the Government of Canada through the Canada Book Fund for our publishing activities.

Published simultaneously in Canada and the United States by

HANCOCK HOUSE PUBLISHERS LTD.
19313 Zero Avenue, Surrey, B.C. Canada V3S 9R9
(604) 538-1114 Fax (604) 538-2262

HANCOCK HOUSE PUBLISHERS
1431 Harrison Avenue, Blaine, WA U.S.A. 98230-5005
(604) 538-1114 Fax (604) 538-2262

Website: **www.hancockhouse.com**
Email: **sales@hancockhouse.com**

Contents

Foreword

Thomas Steenburg

The sasquatch mystery of western Canada has a long, rich history of stories told around hundreds of campfires, wood burning stoves, and folks just relaxing on the porch on a warm summer evening. Stories of this creature have been repeated from generation to generation. Television documentaries, books, radio programs, and of course the internet, remind us every day that the mystery of the sasquatch continues to fascinate and captivate the people of Canada, and remind them of the fact that there is still wilderness out there.

Like most on-going tales, this subject has what most people refer to as its classics. The story of Jacko and his capture near railway tunnel Number 4, just north of Yale, in 1884 is one of the most well-known, and also one of the least documented classic tales of the sasquatch ever told in British Columbia. This story, 126 years later, still intrigues those who believe that such a creature exists, and those who say it cannot possibly exist.

The main question in this tale still persists: who or what was Jacko? Did this story about the capture of such a creature really happen, or were the newspapers at the time just trying to fill in space with spooky tales of wild men outside a frontier town—which in 1884 had a reputation that made Tombstone, Arizona look tame.

Now at last this booklet has been produced concentrating on the story of Jacko. For the first time readers can learn about the circumstances and events that resulted in this tale. They can judge for themselves if the story is a remarkable account of a young sasquatch capture before anybody really knew what they were dealing with (thus missing the opportunity to solve a great mystery, even before the mystery became great) or if was all simply a case of tabloid journalism in its infancy.

— THOMAS STEENBURG

The Jacko Saga

In 1884, an intriguing article appeared in the *Daily Colonist*, a Victoria, British Columbia, newspaper. The article states that a creature, "something of the gorilla type," had been captured near Yale, British Columbia. The creature is described as standing about 4 feet, 7 inches (1.4 m) in height and weighing 127 pounds (57.5 kg). From these details, we might conclude that it was a young sasquatch. The following is an exact reprint of the article, with one exception as indicated. (An image of the actual article is provided in Appendix. A)

What Is It?

A Strange Creature Captured Above Yale

A British Columbia Gorilla

(Correspondence of The Colonist)

Yale, B.C., July 3, 1884.[1] In the immediate vicinity of No. 4 tunnel, situated some twenty miles above this village, are bluffs of rocks which have hitherto been unsurmountable, but on Monday morning last were successfully scaled by Mr. Onderdonk's employees on the regular train from Lytton. Assisted by Mr. Costerton, the British Columbia Express Company's messenger, and a number of gentlemen from Lytton and points east of that place who, after considerable trouble and perilous climbing, succeeded in capturing a creature which may truly be called half man and half beast. "Jacko," as the creature has been called by his captors, is something of the gorilla type standing about four feet seven inches in height and weighing 127 pounds. He has long, black, strong hair and resembles a human being with one exception, his entire body, excepting his hands, (or paws) and feet are covered with glossy hair about one inch long. His fore arm is much longer than a man's fore arm, and he possesses extraordinary strength, as he will take hold of a stick and break it by wrenching or twisting it, which no man living could break in the same way. Since his capture he is very reticent, only occa-

1. The actual newspaper article sows the date as 1882. This appears to be an obvious error that has been corrected in this reprint.

sionally uttering a noise which is half bark and half growl. He is, however, becoming daily more attached to his keeper, Mr. George Tilbury, of this place, who proposes shortly starting for London, England, to exhibit him. His favorite food so far is berries, and he drinks fresh milk with evident relish. By advice of Dr. Hannington raw meats have been withheld from Jacko, as the doctor thinks it would have a tendency to make him savage. The mode of capture was as follows: Ned Austin, the engineer, on coming in sight of the bluff at the eastern end of the No. 4 tunnel saw what he supposed to be a man lying asleep in close proximity of the track, and as quickly as thought blew the signal to apply the breaks. The brakes were instantly applied, and in a few seconds the train was brought to a standstill. At this moment the supposed man sprang up, and uttering a sharp quick bark began to climb the steep bluff. Conductor R. J. Craig and Express Messenger Costerton, followed by the baggageman and brakesmen, jumped from the train and knowing they were some twenty minutes ahead of time gave immediate chase. After five minutes of perilous climbing the then supposed demented Indian was corralled on a projecting shelf of rock where he could neither ascend or descend. The query now was how to capture him alive, which was quickly decided by Mr. Craig, who crawled on his hands and knees until he was about forty feet above the creature. Taking a small piece of loose rock he let it fall and it had the desired effect on rendering poor Jacko incapable of resistance for a time at least. The bell rope was then brought up and Jacko was now lowered to terra firma. After firmly binding him and placing him in the baggage car "off brakes" was sounded and the train started for Yale. At the station a large crowd who had heard of the capture by telephone from Spuzzum Flat were assembled, each one anxious to have the first look at the monstrosity, but they were disappointed, as Jacko had been taken off at the machine shops and placed in charge of his present keeper.

These drawings by Duncan Hopkins show the hapless Jacko beside the railroad trac and then tied up and placed the train's baggage car.

> The question naturally arises, how came the creature where it was first seen by Mr. Austin? From bruises about its head and body, and apparent soreness since its capture, it is supposed that Jacko ventured too near to the edge of the bluff, slipped, fell and lay where found until the sound of the rushing train aroused him. Mr. Thos. White and Mr. Gouin, C.E., as well as Mr. Major, who kept a small store about half a mile west of the tunnel during the past two years, have mentioned having seen a curious creature at different points between Camps 13 and 17, but no attention was paid to their remarks as people came to the conclusion that they had either seen a bear or stray Indian dog. Who can unravel the mystery that now surrounds Jacko? Does he belong to a species hitherto unknown in this part of the continent, or is he really what the train man first thought he was, a crazy Indian?

On July 5, 1884, the article was reprinted in *The Columbian*, another BC paper, and then on July 15, 1884, it was reprinted in the *Manitoba Daily Free Press* under the heading "A Man-Beast: Alleged Capture of a Frontier Creature in B.C." During this time (July 5 to July 15) two newspaper articles (as follows) on the event indicate that the entire story was a hoax. However, we have information provided by a game guide, Chilco Choate, who stated his grandfather was there when this "ape" was brought in and kept at Yale. Whether grandfather Choate actually saw the creature is uncertain (see Appendix B). Next we have a Mrs. Hilary Foskett, who stated that her mother was in Yale at the time (she was about 8 or 9 years old), and remembered stories of the creature. A Dr. Hannington, who is mentioned in the *Daily Colonist* article, was well-known to her. (See Appendix C and E). In this connection, it has been established that all of the people mentioned in the article were real people. Furthermore, sasquatch researcher John Green interviewed Yale resident August Castle in 1958 who, as a young boy, remembered hearing about Jacko[2] (see Appendix D and E). Lastly, Ellen Neal, a noted First Nations artist, was told by Chief August Jack Khahtsahlano that a creature of this nature did reach Burrard Inlet in Vancouver and was exhibited there. John Green

2. The only marginal evidence we have that someone did see Jacko is this statement in Ivan Sanderson's book, *Abominable Snowmen: Legend Come to Life* (27, 28): "a reporter in 1946 interviewed an old gentleman in Lytton, B.C., who remembered seeing it."

states that he was told that Chief Khahtsahlano actually saw a creature (i.e., a sasquatch-like creature) on display in 1884. It is reasonable to assume that this was Jacko.

The two BC newspaper articles that imply the entire story was a hoax are reprinted here:

Chief August Jack Khahtsahlano

The What is It
Yale, British Columbia, July 9, 1884

THE WHAT IS IT is the subject of conversation in town this evening. How the story originated, and by whom, is hard for one to conjecture. Absurdity is written on the face of it. The fact of the matter is, that no such animal was caught, and how *The Colonist* was duped in such a manner, and by such a story is strange; and stranger still when *The Columbian* reproduced it in that paper. The "train" of circumstances connected with the discovery of "Jacko" and the disposal of same was, and still is, a mystery. (*Mainland Guardian*)

The Wild Man
Yale, British Columbia, July 12 1884

Last Tuesday [July 8] it was reported that the wild man, said to have been captured at Yale, had been sent to this city and might be seen at the gaol [jail]. A rush of citizens instantly took place, and it is reported that not fewer than 200 impatiently begged admission into the skookum house. The only wild man visible was Mr. Moresby, governor of the goal, who completely exhausted his patience answering enquires from the sold visitors. (*The Columbian*)

There is no indication that the *Daily Colonist* defended their original story. This fact, perhaps, is a major indication that the story was fabricated. However, it is unusual that two other newspapers featured the story if the *Daily Colonist* knew it was hoaxed and still allowed them to use it. Whatever the case, the story is tightly woven into sasquatch/bigfoot lore and has become a part of the history of Yale.

This postcard showing Yale in the 1880s gives one a good idea of how the little town appeared at that time. Additional photographs are provided in Appendix F.

The town of Yale in the 1880s.

Morse Lister of Yale. His grandfather, Dr. Garnet Morse, worked with Sir Robert Baden-Powell to bring the Boy Scouts to Canada.

Today, there is absolutely nothing left of the old town. Yale consists of a number of private houses and a few stores on the main highway that cuts through the middle of the town.

I have often wondered where the photograph of the steamship was taken, and recently went up to Yale with Barry Blount to see if we could find the spot. A resident, Morse Lister, was kind enough to explain things to us.

After a trek down to the Fraser River, Morse showed us what is left of the dock where the old steamboat is moored. I went up the bank and found a wide trail which was once the main road seen in the postcard. After some 120 years, the entire area is now

Three crumbling cement structures, with a fourth in the bush line, mark what is left of the docking facility.

Barry Blount holds an iron ring mounted with an eye bolt in a huge boulder at the water's edge. It was, and still is, used as a boat tie. We found this artifact on a subsequent trip to the site.

heavily overgrown. What is left of the old dock is seen in the photograph. The trees at the back of the photograph conceal the wide trail that was once the main road in the town.

Below are shots of the trail. Beyond the trail to the right there is a bank and then Front Street, a paved road. Rotting wood seen here was possibly part of one of the original structures.

The trail above the beach that once formed the main road in Yale.The fourth cement dock structure is seen on the left.

Rotting wood seen here may be what is left of one of the original buildings that once lined the main road in Yale.

I was taken by Morse to a large block of cement, as seen in the next images, at about the end of the trail, and told that it was part of the Yale jail. Morse said it was originally further up the bank and that he had pushed it down to its present location.

Cement block said to have been a part of the original Yale jail.

As can be seen, the block is hollow. I put my camera inside and flashed a picture as seen here.

Inside the cement block said to be from the Yale jail. It appears to be framed in plywood, which was invented in the 1850s, so that is probably what it is.

We are told in the story of Jacko that he was taken to the Yale jail, but this was denied by the jail keeper. There is, however, another "Yale jail." It is displayed at a shop a mile or so down the main highway to the west of the town. It is simply a large metal box with two bunks inside, as seen in the following photographs. I would certainly pity anyone, man or sasquatch, who was locked up in that thing.

Author beside the other Yale jail, now located outside the town.

The nagging question as to whether or not the story of Jacko as detailed in the *Daily Colonist* newspaper report could be true has plagued sasquatch researchers since the 1950s. Although it is well written and very detailed, there are serious discrepancies.

Perhaps the main discrepancy is that the article is dated 1882 whereas the date for the paper is 1884. I have assumed that this was a simple typesetting error. Nevertheless, it is still bothering. To even consider that the event actually took place in 1882 and was not published by the *Daily Colonist* until two years later does not make sense unless, of course, the article was intended to be a hoax.

We then have the actual time frame from the date of the event to the date the article was published. From my research, telegraph to Vancouver Island was not in place until 1898, and telephone service not until 1927. This means that the article had to be hand delivered to the newspaper for a part of its journey.

We are told in the article that the event (capture of Jacko) took place on Monday, June 30. As the article was in the Friday morning (July 4) edition of the *Daily Colonist*, then it had to be in hand by the paper on Thursday, July 3 at the very latest, which is the date shown on the article. This means that whoever wrote the article (the correspondent) had just two days to research all the information and get the article to Victoria, BC, the location of the *Daily Colonist* (118 miles as the crow flies).

There was telephone and telegraph service to New Westminster and Vancouver. One source says that the story was a "telegraph dispatch," so this being the case, it was telegraphed to either of these cities and then hand delivered to the newspaper in Victoria. It was certainly a very long telegraph, and therefore very expensive. Whatever the case (telegraph or telephone) it is feasible that the correspondent could have gotten the information to Victoria by the morning of July 3.

Nevertheless, the tight time frame is still bothering. I don't think news of this nature would have been considered urgent, and thus would not warrant the expense to rush it to Victoria.

The inference here appears to be that the article was definitely a hoax, written by someone perhaps in 1882 and sent to the newspaper. We can reason that it was probably filed away and then pulled and published two years later as "filler." Naturally, its publication would stir things up in Yale and all sorts of residential speculation would result.

Sasquatch researcher John Green (2010), left, and René Dahinden (1993), initially investigated the Jacko incident in 1958, using information provided about the incident (i.e., verbal). The actual (physical) *Daily Colonist* newspaper article, which was on microfilm, was not uncovered until 1974 and it resulted in more intensive research.

However, at the time John Green and René Dahinden investigated the Jacko story in 1958, they worked with Yale historian August "Gus" Milliken who knew nothing about the incident. Milliken achieved notoriety after he uncovered First Nations artifacts at a spot about two and one-half miles north of Yale. The resulting archaeological dig produced evidence that people had lived there over 9,000 years ago. The dig was named the Milliken Site, and is one of the oldest in British Columbia.

Certainly, Milliken was an astute and meticulous historian, so it is indeed a bit odd that he never saw the *Daily Colonist* article or came across anything related to the story (diaries, letters, notes, and so forth). Although the story itself may not be true, the fact that it was published is definitely true. All I can think of here is that everyone at the time knew, or found out, the story was a hoax and simply disregarded it (including the local historian, whomever he or she was).

We then have the second-hand statement that Chief Khahtsahlano actually saw a creature of Jacko's description on display at Burrard Inlet in Vancouver in 1884. As this is the only reference we have of the "showing," it becomes suspicious. If it were true, one would think there should have been a written account of the incident somewhere. If Jacko was important enough for an immediate *Daily Colonist* newspaper article, then someone should have seen and written about him when he was displayed in Vancouver—not necessarily, but logically.

Next we have the discrepancy as to actual location where Jacko was captured. The article states: "In the immediate vicinity of No. 4 tunnel, situated some 20 miles above this village [Yale]." The No. 4 tunnel is only about 2 miles from Yale. The following images show both the tunnel and a spot about 20 miles from Yale.

The No. 4 railway tunnel (top), about 2 miles from Yale, and a spot about 20 miles from Yale. Both are given as the site of Jacko's capture.

I have provided a map in Appendix G along with what I believe is an explanation for the discrepancy.

Another point that has bothered researchers is the way in which Jacko was captured. In the first place, that men from the train were able to corner Jacko is "pushing the envelope." Jacko, if real, was wild and would likely have been able to out-climb and out-smart his pursuers. As to one of the men being able to drop a "small piece of rock" from 40 feet and knock out Jacko, it is possible, but a long shot.

Given the story of Jacko is true, the burning question is, what happened to him? The last we know is that he was shipped in a cage to the east or to England to be used in a sideshow, but he apparently never arrived at either destination.

There are, however, two unsubstantiated diversions in the story. Firstly, in 1972 Annie York (1904–1991) a Spuzzum First Nations elder and historian, said she recalled that her father (or grandfather) had buried a sasquatch that had been captured (lassoed by train workers) at a railway tunnel above Spuzzum (not Tunnel No. 4 above Yale). We are told that when it was lassoed the creature's neck was broken and it died. I need to mention here that Annie York had been interviewed by John Green in 1958 and she made no mention of this incident. Secondly, after possibly being displayed in Vancouver,[3] Jacko was said to have been taken up Burrard Inlet where people cut off his hair to see what it was like underneath. He then died.

In the 1970s, Dr. Grover Krantz followed up on the possibility of rail shipment of Jacko across the United States. He was informed that there was a newspaper account of a strange animal arriving from the west in Duluth, Minnesota in 1884. Two people who did not know each other provided this information, so Krantz thought it might be credible. He checked the surviving Duluth newspapers from the 1880s, but was unable to find anything. He then visited three towns on the Canadian Pacific Railroad to see if there was any evidence of Jacko passing through. He talked to the oldest "functioning" people in the towns (but all under 90 years old) who had lived there all their lives, and asked them if they had heard any Jacko stories passed down from relatives or friends who were alive in the 1880s. None of these people had any recollections, and they

3. We can only assume that the creature possibly displayed in Vancouver was Jacko.

were sure that if the creature had passed through, they would have heard about it. However, there is some interesting speculation.

Across the continent, also in the year 1884, the Barnum & Bailey Circus presented in New York City Jo-Jo the Dog-Faced Boy, a sixteen-year-old youth covered in long hair. Jo-Jo, whose actual name was Fedor Jeftichew (b. 1868), was alleged to have been found in Russia along with his father, who was also covered in hair. Jo-Jo has been coincidentally connected with Jacko. It has been reasoned that Jacko may have been purchased in the United States by circus man P. T. Barnum and billed for a sideshow, but died before he could be exhibited. Barnum thereupon quickly found a replacement—Jo-Jo. From what Krantz learned, circus advertising material created in 1884 showing a hairy creature does not appear to show Jo-Jo. This material was said to have been replaced with an ad showing an actual photograph of Jo-Jo taken in 1885. To my knowledge, the original advertising material has never been found.

When the Barnum & Bailey collection was purchased by a research organization in the eastern United States about three years ago, it posted the contents of the collection to their website. Documents were separated into "drawers," and I saw that one drawer had the title "Jacko." I contacted the curator and asked her to inform me what was in the drawer. A short time later, she got back to me and informed that the drawer was empty. I have not been able to find any reference to the word "Jacko" in any other connection to Barnum & Bailey (e.g., a clown, or something like that).

Spuzzum (or Spuzzum Flat)

We are told in the *Daily Colonist* article that people in Yale were informed of Jacko's capture by telephone from Spuzzum. This, of course, leads us to believe that Jacko was seen in Spuzzum. For this to have been the case, then Jacko must have been captured further to the north, assumably the 20 miles from Yale location. I have previously discussed the location discrepancy and presented a map in Appendix G. Whatever the case, I will share our research on Spuzzum.

The little settlement is about 8 miles by road east of Yale. At this time, fewer than 70 people reside in Spuzzum. Generally speaking, it is a scattering of dwellings (old houses and portable buildings/ trailers) with a railway running through the middle. The main highway (No. 1) can be heard from the settlement, and vehicles can be seen whizzing by through the trees. The settlement was immortalized in the early 1980s by the band Six Cylinder in a song with the refrain, "If you haven't been to Spuzzum, you ain't been anywhere."

We saw a man working on his trailer and stopped to ask him questions. He came over to our SUV and we proceeded to discuss Spuzzum with him. His name is Joe Haslam, certainly a great guy, who has lived in Spuzzum since he was 14 years old (1957). He was originally from Vancouver Island, and as a boy, was sort of "adopted" by the Spuzzum First Nations people, whose reservation borders Spuzzum to the south.

Joe Haslam

We asked Joe if he had heard of the Jacko, story, which he had. We asked if he believed it, and he said it could definitely be true. He went on tell us that all his friends were the Spuzzum First Nations people and he had heard from them of sasquatch encounters up in the woods to the east. Joe has never seen one of the creatures, but he is certainly a believer.

The train did stop in Spuzzum, and Joe showed us what is left of the loading platform (seen below), which was almost in front of his property.

Train dock in Spuzzum.

House that belonged to John MacInnes, Joe Haslam's neighbor.

Joe remembered that when he arrived in Spuzzum in 1957, there was an old public telephone—likely the only telephone in the settlement.

As we stood talking, I noticed a little old ramshackle house (as shown on the right), probably 100 feet or more down the road. Joe told us the house belonged to John MacInnes, who died about 30 years ago. He was in his 70s and died in the old house while playing solitaire.

MacInnes was Spuzzum First Nations, so likely lived in the area all his life. I mused that his father and mother were probably there in the 1880s. More images of Spuzzum are in Appendix H.

Author (left) with Chief Mel Bobb of the Spuzzum First Nations people. Chief Bobb mentioned that his people have talked of the sasquatch.

The Machine Shops

In the *Daily Colonist* article we learn that Jacko had been taken off at the "machine shops." These shops were located about three-quarters of a mile (1.2 km) north of Yale (actual town). It appears that what is now Toll Road was once the original railway line into Yale. The shops were at a point where a tunnel had to be built to provide a more direct (straight) line into the town. For anyone in Yale to have actually seen Jacko at the machine shops, they would have needed to travel the distance (3/4 mile) beside or on the railway line. It is evident few, if any, people other than Dr. Hannington, did this. The following photographs show the machine shop and the location as it presently appears.

The Yale machine shops. What is seen on the left that appears like a tunnel is actually a tree. The tunnel was not yet completed.

The machine shops' location as it presently appears. You can see the tunnel on the left. The bank on the right appears to have eroded considerably over the last 126 years.

An Interesting Side Track

In looking at the full page of the *Daily Colonist* newspaper, I was amused to see a Chinese Petition (shown in Appendix I) directly beside the Jacko article. It provides a sense of some mind-sets back in the 1880s.

This petition, and I am sure others like it, led to the Canadian Government Chinese Head Tax and the Chinese Exclusions Act.

There were thousands of Chinese railroad workers on the Canadian Pacific sea-to-sea railroad, and one would think that some of them would have had sightings of unusual creatures as the railroad pushed from east to west. Unfortunately, the diabolical relationship many non-Chinese people chose to maintain with the workers apparently resulted in total silence. I have never run across a sasquatch-related report of this nature.

Chinese railway workers (Coolies). Photograph taken about 1884.

Nevertheless, one might wonder as to the reaction of Chinese workers at the time if it were proven that some sort of ape-man inhabited North America's forests. Despite protests, these workers were urgently needed, so there would likely have been a move to "put a lid on things" just to be on the safe side.

There is a monument (see Appendix J) outside the museum at Yale which pays tribute to the Chinese railroad workers (Coolies) and effectively apologizes for the inhumane treatment of these workers.

In 2006, Prime Minister Stephen Harper made a formal declaration of apology to Chinese people for the way their relatives had been treated.

Concluding Remarks

Although the Jacko story now carries a "red flag" in the annuls of sasquatch lore, it is nonetheless a part of Yale's history. It might be noted that intensive study of the sasquatch question certainly indicates that an event of this nature could have happened, and may indeed happen in the future.

One thing I must admit, however, is that when Barry and I found the photo of the machine shops (where Jacko was dropped off) and then traced their location, such added a measure of credibility to the Jacko story. The rumor that the creature had been transferred to the Yale jail could have been started to put people off track. Keep in mind there is no mention of the jail in the *Daily Colonist* article. Mr Tilbury could have taken Jacko to his residence where he was seen by Dr. Hannington. After a few days, Tilbury could have quietly sent or taken Jacko to the coast (before the article appeared in the *Daily Colonist*).

Given we can forgive the discrepancies in the Jacko story, we are still left with the question as to why the *Daily Colonist* (to our knowledge) did not react when other newspapers inferred or called the story a hoax. There is also the question as to why (again to our knowledge) the people mentioned in the article (especially Dr. Hannington) did not complain about their names being shown in connection with a hoax.

Pure speculation, I admit, but what we have here could be what we now call a "gag order." Remarkably, this sort of thing does appear to happen with regard to unexplained phenomena.

The question as to why there is no information on the incident in the public archives of the Yale Museum has indeed troubled researchers. However, the museum people are still working through their current collection, and as time goes by certainly additional documentation will find its way to the museum.

I think the final episode in the Jacko saga is yet to be written, and until then I don't think the story should be written off as a proven hoax.

General British Columbia Sasquatch Facts: The first BC report of what might have been one of the creatures happened in 1792. Over the last 100 years, more than 360 sasquatch-related incidents (sightings and large footprint findings) have been documented. As the ratio of non-reported incidents to reported incidents is believed to be about 8 to 1, then the true total figure is likely close to 3,000.

Sightings of the creature in British Columbia continue, and researchers believe it is only a matter of time before the sasquatch issue is fully resolved. A map showing sightings in British Columbia is provided in Appendix K.

Appendices

Daily Colonist Newspaper Page

Daily Colonist.

FRIDAY MORNING, JULY 4th, 1884.

SHIPPING INTELLIGENCE.

PORT OF VICTORIA, BRITISH COLUMBIA.

ENTERED.

July 3—Str E. P. Ritbet, Nanaimo
Str Yosemite, New Westminster
Sch Annie Beck, Sooke
Str Queen of the Pacific, Pt Townsend
Str Olympian, Pt Townsend

CLEARED.

July 3—Str Dolphin, Sooke
Str R. P. Rithet, Nanaimo
Str Yosemite, New Westminster
Str Queen of the Pacific, San Francisco
Str Olympian, Pt Townsend

PASSENGERS.

Per str OLYMPIAN from Puget Sound—[illegible]

Per ss QUEEN OF THE PACIFIC, fm San Fran.—[illegible]

Per str YOSEMITE, from New Westminster—[illegible] ... and 4 Chinese.

CONSIGNEES.

Per str OLYMPIAN, fm Puget Sound—C Strouss, [illegible], Neufelder & Ross.

New Westminster.

(Special to The Colonist.)

NEW WESTMINSTER, July 3.

Custom receipts at this office the fiscal year ending 30th of June, 1884, was $94,-728 56, a decrease of $16,667.02 as compared with the previous year, decrease caused by the increased imports from eastern provinces.

Constable Wiggins was brought up today before the police magistrate charged with assaulting one Richardson and was fined $75 and costs or three months' imprisonment.

Chinese Petition.

The following petition has been got up for general circulation and signature in this province, and is to be presented at the next meeting of the parliament of Canada, and we are informed that a similar petition to the Dominion parliament is to be circulated through the eastern provinces for signature:

To the Honorable the House of Commons in Parliament Assembled: The petition of the undersigned residents of British Columbia, humbly sheweth—

That we view with great alarm the thousands of Chinese Coolies continually arriving in the province of British Columbia, and while we regard this country as free to every man seeking an opportunity to better his condition in life, we deem it but just and right that those who come to our shores shall not be such as to work injury to the moral and material welfare of the Dominion. That the unlimited admission of the race of people known as Chinese Coolies does work such injury to our country, for the following reasons: That they do not come to make a home or settle in the country, or to add to the country's wealth; but to prey upon our natural resources, and take what they earn out of the country. That they are leprous in blood and unclean in habits. That they are destructive of the means by which the white mechanical and laboring classes earn a living wage. That there are immoral practices, debasing habits and contagious diseases, peculiar to this people, which they have already introduced to an alarming extent upon the continent, and against which we have a right to defend ourselves and our children. We therefore humbly pray for the enactment of such laws as will prohibit any farther introduction of this undesirable class of people into any part of the Dominion of Canada. And as in duty bound your petitioners will ever pray.

The Ben Cotton Company.

Last night the Ben Cotton Company opened their week's engagement at the Philharmonic to a crowded house. The play presented was a comedy drama in a prologue and three acts, and is a succession of thrilling pictures, interspersed with fun and frolic, and the audience were kept in a constant delight from the prologue to the fall of the curtain. Mr Cotton, as a delineator of negro character, has few equals on the stage, and that of "Bob the Bootblack," affords fine scope for his powers. His pretty little daughter, Idalene, captivated the house from the first. Her imitations are original and the charming way

WHAT IS IT?

A STRANGE CREATURE CAPTURED ABOVE YALE.

A British Columbia Gorilla.

(Correspondence of The Colonist.)

YALE, B. C., July 3rd, 1882.

In the immediate vicinity of No. 4 tunnel, situated some twenty miles above this village, are bluffs of rock which have hitherto been unsurmountable, but on Monday morning last were successfully scaled by Mr. Onderdonk's employees on the regular train from Lytton. Assisted by Mr. Costerton, the British Columbia Express Company's messenger, and a number of gentlemen from Lytton and points east of that place who, after considerable trouble and perilous climbing, succeeded in capturing a creature which may truly be called half man and half beast. "Jacko," as the creature has been called by his capturers, is something of the gorilla type standing about four feet seven inches in height and weighing 127 pounds. He has long, black, strong hair and resembles a human being with one exception, his entire body, excepting his hands, (or paws) and feet are covered with glossy hair about one inch long. His fore arm is much longer than a man's fore arm, and he possesses extraordinary strength, as he will take hold of a stick and break it by wrenching or twisting it, which no man living could break in the same way. Since his capture he is very reticent, only occasionally uttering a noise which is half bark and half growl. He is, however, becoming daily more attached to his keeper, Mr. George Tilbury, of this place, who proposes shortly starting for London, England, to exhibit him. His favorite food so far is berries, and he drinks fresh milk with evident relish. By advice of Dr. Hannington raw meats have been withheld from Jacko, as the doctor thinks it would have a tendency to make him savage. The mode of capture was as follows: Ned Austin, the engineer, on coming in sight of the bluff at the eastern end of the No. 4 tunnel saw what he supposed to be a man lying asleep in close proximity to the track, and as quick as thought blew the signal to apply the brakes. The brakes were instantly applied, and in a few seconds the train was brought to a standstill. At this moment the supposed man sprang up, and uttering a sharp quick bark began to climb the steep bluff. Conductor R. J. Craig and Express Messenger Costerton, followed by the baggageman and brakesmen, jumped from the train and knowing they were some twenty minutes ahead of time immediately gave chase. After five minutes of perilous climbing the then supposed demented Indian was corralled on a projecting shelf of rock where he could neither ascend nor descend. The query now was how to capture him alive, which was quickly decided by Mr Craig, who crawled on his hands and knees until he was about forty feet above the creature. Taking a small piece of loose rock he let it fall and it had the desired effect of rendering poor Jacko incapable of resistance for a time at least. The bell rope was then brought up and Jacko was now lowered to terra firma. After firmly binding him and placing him in the baggage car "off brakes" was sounded and the train started for Yale. At the station a large crowd who had heard of the capture by telephone from Spuzzum Flat were assembled, each one anxious to have the first look at the monstrosity, but they were disappointed, as Jacko had been taken off at the machine shops and placed in charge of his present keeper.

The question naturally arises, how came the creature where it was first seen by Mr. Austin? From bruises about its head and body, and apparent soreness since its capture, it is supposed that Jacko ventured too near the edge of the bluff, slipped, fell and lay where found until the sound of the rushing train aroused him. Mr. Thos. White and Mr. Gouin, C. E., as well as Mr. Major, who kept a small store about half a mile west of the tunnel during the past two years, have mentioned having seen a curious creature at different points between Camps 13 and 17, but no attention was paid to their remarks as people came to the conclusion that they had either seen a bear or stray Indian dog. Who can unravel the mystery that now surrounds Jacko? Does he belong to a species hitherto unknown in this part of the continent, or is he really what the train men first thought he was, a crazy Indian?

H. M. S. Constance "At Home."

Although the hospitable intentions of Captain Doughty and his officers were frustrated last week by the inclemency of the

What Some People Say.

That J. B. Ferguson & Co. are sole agents for Fairchild's Gold Pens for the province of British Columbia and have just received a large stock. Call and see them. •

That a fresh lot of San Juan strawberries and a lot of chickens (broilers) were received this morning at Beauchamp's, Yates street. •

A large stock of Seasides just received at J. B. Ferguson & Co.'s new book store—third door south of post office. •

A few more Lawn Tennis Sets at J. B. Ferguson & Co.'s just opened. •

That the King Tye Co. are leading manufacturers and importers of clothing, groceries, etc. •

That for neatness and cheapness in repairing you should go to Lavinu Bass.' Boot and Shoe Store, Government street, between Broughton and Courtenay. They make repairing a specialty. •

That Clothing, Hats and Underclothing are very cheap at W. & J. Wilson's. Look at their prices and you will see that their stock is the best value in Victoria. •

That the new Customs Excise Tariff with its refreshing, onerous facts, has arrived at last. T. N. HIBBEN & Co.

That employers can be supplied with every class of labor by applying to Knights of Labor Bureau, Yates Street. •

That you should go to John Weiler's, Fort street, and look at the beautiful Walnut and Ash Furniture, which has just arrived from the East. Do not miss the chance. •

That for
best value
in clothing,
furnishing goods,
trunks, &c.,
go to the
Mechanics' Store,
Johnson street,
W. G. CAMERON.

That the celebrated Val Blatzs Milwaukee is for sale at the "Senate." •

That THOMAS NICHOLSON, Family Grocer, corner Douglas and Johnson streets, has a new stock of Groceries and Provisions, Lard, Hams and Bacon, packed by the famous house of Logus and Lierenger, Oregon city. Goods delivered. Fine Wines and Liquors. Prices low. *3m

"PATRONIZE HOME INDUSTRY."—Rubber Stamps of all kinds made to order. R. T. WILLIAMS, Broad street. •

That you should go to J. B. Ferguson & Co's for Base Ball and Cricket goods. •

That the Leading clothing and furnishing House in the Province is the
Scotch House,
A. McLEAN & Co.
Established
1862.

3m•

That a seven roomed house, pleasantly situated on the water front on Dallas road, is to rent. Rent $20 per month; apply to L. Lowenberg. •

Kamloops.

A gentleman who arrived direct from Kamloops via the C. P. R. and Yosemite yesterday furnishes us with the news that a welcome rain fell on the 23rd and 24th of June doing immense benefit to the standing crops. The farm of Mr. Lumby, in the Spallumcheen valley, was in splendid condition and that gentleman estimates that wheat and barley would average one ton to the acre. Stock was also looking well and after the rain there was plenty of pasture. Everyone was feeling jubilant over the prospect for a good average harvest.

Railway construction is going on in an energetic manner and the line is now graded as far as Spence's Bridge while trains are running for considerable distance beyond Lytton.

Granville Public School.

The annual examination of the above school took place on Sunday. The room was tastefully decorated with evergreens and together with the childrens' happy faces presented a bright appearance. In the afternoon the trustees and parents assembled when the pupils were examined in geometry, spelling, grammar, history and reading, all showing great proficiency and reflecting much credit on their teacher, Miss Gardiner. A prize offered for the best English composition was won by Miss Lizzie Alexander, the young lady reading an essay that would have done credit to many a high school pupil. A number of other prizes were distributed and after singing the National Anthem the school closed for the summer vacation.

C. P. N. Co., Limited.

The public are requested to note the change in time-table of C. P. N. Co.'s

Image of most of the full page in the *Daily Colonist* showing the Jacko article.

B Letter from Chilco Choate

Letter written to Dr. Grover Krantz in 1970 by Chilco Choate, a game guide at Clinton, BC

Received your letter re my grandfather's experience with the sasquatch which was captured near Yale, B.C. in the last century. The case you mentioned is the same to be sure.

I heard this story from my father (who is now getting on and his memory is getting a little vague most of the time). When he first told us the story many years ago it was still quite clear in his mind and this is how it went.

My grandfather was the B. & B. engineer (buildings & bridges) for the C.P.R. when it was being built west of Revelstoke. There is still a small train stop near Yale that was name after him (Choate, B.C.). After the C.P.R. was built he became a circuit judge for the County Court of Yale, although I don't know just how long he was actually a judge. Anyway, he was there when the "ape" was brought in and kept at Yale. "Ape" was the word my dad says his dad called the captive. The ape was kept in Yale until the owner loaded it crated onto a train heading east. The story goes that he was taking it to London, Eng. to set up a side show with it and make his fortune. This was the end of the story as nobody heard of either of them again. It was either my grandfather's opinion or dad's opinion that the ape must have died on the trip and was probably disposed of in any-way possible. Personally I would imagine it probably died at sea and would have simply been thrown overboard.

Note: Whether grandfather Choate actually saw the creature is uncertain. However, I think he would have said so if such was the case. Also it should be noted that the railway going to the east was not fully completed in eastern BC, so what the letter states is not correct. However, Jacko could have been sent to Vancouver, and from there to the US and subsequently by rail to the east.

Letter from Mrs. Hilary Foskett

C

Letter written to John Green by Mrs. Hilary Foskett of Ucluelet, BC, c. 1970

Did I mention before that my mother, Adela Bastin, was educated in Yale at All Hallows in the West, a school run by Anglican nuns from All Hallows in Ditchingham, England? The school was of course a boarding school and accepted pupils from all over.

When the stories of the "Yeti" and Sasquatch appeared in the press, Mother recalled stories of Jacko at Yale. She was probably eight or nine when she started school there and local inhabitants were still talking about the "wild man," and the good sisters at the school took care in shepherding the pupils from school to chapel or church. In spite of this local "fear" in her later years at the school, Mother climbed Mt. Leakey behind Yale with a group of local young people. Until well in her eighties she could recall Yale days in detail, but before her death at 93 a year ago, her memory was slipping. The Dr. Harrington *[see Note]* referred to was well-known to mother and her sisters…

Note: In the article the name is "Hannington."

D Vancouver Sun Article – October 29, 1958

Agassiz Folk Examine Caves

SASQUATCH HUNT MOVES FROM HARRISON TO FRASER CANYON

By A. C. MILLIKEN

The Sasquatch hunt has now moved from Harrison Lake area to Yale. Last Sunday Mr. and Mrs. John Green of Agassiz, accompanied by Mr. and Mrs. Rene Dahinden, visited the Yale area and accompanied by friends from Yale, examined a cave reported to have been the home of a Sasquatch.

Mr. Green, who is publisher of the Agassiz Advance, has done considerable research in this connection. His attention was drawn to this area by a news story carried in the Victoria Colonist under the date of July 3, 1884.

Briefly, the story reports that a creature, described as half man and half beast, was captured in the vicinity of No. 4 tunnel above Yale. It is described as resembling a human being with but one exception. His entire body is covered with black glossy hair about once inch long, except the hands (or paws) and feet. It possesses extraordinary strength and occasionally utters a noise described as half bark and half growl.

The Colonist article gives a dramatic description of the capture including the names of the participants, who have been found to be persons living in Yale at the time, and employed on the railroad construction. All were men of importance and holding responsible positions, one being the engineer in charge.

Mr. Green interviewed August Castle, dean of oldtimers in this area. Mr. Castle was a young boy at the time and does not remember the incident but he does have a recollection of hearing later that such a "thing" had been caught.

Annie York of Spuzzum was also interviewed. While Miss York does not have the age qualifications to verify the story, her wealth of knowledge of early incidents as told to her by the old people years ago, yielded no clue in this connection. However, Miss York related many stories of reported sighting of Sasquatch at different points in Canyon area.

A preliminary investigation of the above mentioned cave leads one to believe that it has been inhabited at one time by some form of life. Some osteological material was found but not sufficient to prove conclusively what it was.

Note: John Green would later (1978) write: "Some years ago I talked to an old man named August Castle who was said to have been at Yale all his life. He said that he was a small child in 1884 and lived in the Indian section of the community. He did not see Jacko, because it was the white man who had the animal, but he remembered the excitement about it." (*Sasquatch: The Apes Among Us*, p. 86. See Appendix E for evidence of the Castle family.)

MIDWIVES (FROM NOVEMBER 7, 1880 ON)

[illegible]	ACCOUCHEUR	CHILD
1881	Dr. Mclean	Bears
	Mrs. James	Wardle
	Dr. Hanington	Lewis
	" "	" "
	" "	McDonald
	" "	Oldman
1882	Mrs. Allway	Thrift
	Hanington	VanVolenburgh
	Frickelton	Ralyar
	Mrs. Ward	Horth
	Hanington	Taylor
	Emily H. Ward	Dodd (twins)
1883	Hanington	Teague
	Hanington	Michell
	Hanington	White
	Frickelton	McQuarrie
	Frickelton	McQuarrie
	Hanington	Campbell
	Dr. E.B.H. Hauvey	McKay
	Hanington	Dunn
1884	Hanington	Irwin
1885	Hanington	Dawzy
	Mrs. McQuarrie	Bossi
	Hanington	Tuttle
1886	Mrs. McQuarrie	Brown
	Mrs. C. Fraser	McQuarrie
	Dr. Cooper	McDougal
	Mrs. Lovelace	Michell
1887	Mrs. James FRaser	Johnson
1888	"Husband present"	McQuarrie
1891	Lily McLinden	Creighton
	Lily McLinden	Castle
	Mrs. McQuarrie	da. of Joseph Corrigan, hotel keeper in Hope
1893	Dr. Williams, FRCS	Flann
	Dr. Williams	Flann
	Mrs. McQuarrie	Corrigan
1894	Dr. J.R. Williams	Johnson
	Dr. J.R. Williams	Harris
1896	"Her mother, an Indian women"	Blachford
1897	Mrs. Gillard	Campbell
1881	Hanington	Oppenheimer
1898	Mrs. McQuarrie	Hantz (Spuzzum)
1901	Mrs. McQuarrie	Davenport (Hope Station)

A gravestone in the Yale cemetery marks the resting place of Robert Castle who died in 1886. This was probably August Castle's grandfather. The stone also shows an Alph Castle who died when only 6 months old. A complete listing of all tombstone information in the Yale cemetery is available at the Yale Museum. No people shown in the *Daily Colonist* article who were involved in the Jacko incident were on the list. This, however, is naturally due to the fact that all, save Dr. Hannington, worked on the train and probably lived elsewhere. Dr. Hannington was evidently buried elsewhere, or was cremated. Nevertheless, that he did live and work in Yale is fully confirmed by the above document that shows him and others who performed midwife service in Yale. It might be noted that his name has only one "n" (Hanington) in the document, which is the correct spelling. His full name was Dr. Ernest Hanington.

F Yale — Additional Photographs

The main street in Yale in the 1880s.

Some of the "boys" playing poker. Yale was a wide open mining town, created upon the discovery of gold (1858) in British Columbia. There were some 9,000 people working between Yale and Lytton (about 47 miles north) in the early 1880s. Few, if any, of these people were interested in anything besides gold and making money. Even the medical doctor who we are told saw Jacko did not apparently realize the scientific significance of the creature.

St. John the Divine Church at Yale (construction completed in 1863) is shown above as it appeared in 1883. It is the second oldest church on mainland British Columbia. I don't know who the people are in the photograph, but it is probably a safe guess that they were in Yale when the alleged Jacko incident took place.

The church underwent extensive renovations in 1953, and the adjacent photograph shows how it appears today.

G Map of Yale Area

This map shows the locations as they pertain to the Jacko story. It appears more logical that the capture of Jacko occurred at the 20 miles from Yale point because we are told people were informed of the incident by telephone from Spuzzum. However, this detail may have been a misunderstanding—the call probably came from the machine shops. Also, in my opinion, the 20 miles figure should have been a 2 (typeset error). If the incident occurred at the 20 mile spot, it would have been stated as about 1 mile south of Boston Bar. (Measurements are approximate railway miles.)

Spuzzum (Spuzzum Flat) — Additional Photographs

H

The railway lines that run through Spuzzum, facing north. The tracks on the left do not appear to be used. They "border" the loading platform (previously shown) in what might be said is in about the center of the settlement.

I believe these buildings are the oldest in Spuzzum. On the left is a large residence. The main (and essentially only) road passes by in front. Beyond the road is a strip of unkempt land that borders the railroad tracks. On the right is what was once the general store, and is now used strictly as a residence. It is located on the other side of the railroad tracks on the same road that winds back in the opposite direction. The old store is visible through the trees from the main part of the settlement.

I

Chinese Petition.

The following petition has been got up for general circulation and signature in this province, and is to be presented at the next meeting of the parliament of Canada, and we are informed that a similar petition to the Dominion parliament is to be circulated through the eastern provinces for signature:

To the Honorable the House of Commons in Parliament Assembled: The petition of the undersigned residents of British Columbia, humbly shewth—

That we view with great alarm the thousands of Chinese Coolies continually arriving in the province of British Columbia, and while we regard this country as free to every man seeking an opportunity to better his condition in life, we deem it but just and right that those who come to our shores shall not be such as to work injury to the moral and material welfare of the Dominion. That the unlimited admission of the race of people known as Chinese Coolies does work such injury to our country, for the following reasons: That they do not come to make a home or settle in the country, or to add to the country's wealth; but to prey upon our natural resources, and take what they earn out of the country. That they are leprous in blood and unclean in habits. That they are destructive of the means by which the white mechanical and laboring classes earn a living wage. That there are immoral practices, debasing habits and contagious diseases, peculiar to this people, which they have already introduced to an alarming extent upon the continent, and against which we have a right to defend ourselves and our children. We therefore humbly pray for the enactment of such laws as will prohibit any further introduction of this undesirable class of people into any part of the Dominion of Canada. And as in duty bound your petitioners will ever pray.

Chinese Construction Workers' Monument

The monument to Chinese construction workers on the Pacific Railway, on display at the Yale museum. An enlargement of the plaque is shown below. I was told by Jake Baerg of the Yale Museum that the Chinese writing at the bottom is depicted incorrectly — it reads horizontally rather than vertically.

CHINESE CONSTRUCTION WORKERS
ON THE PACIFIC RAILWAY
TRAVAILLEURS CHINOIS DU CHEMIN
DE FER DU PACIFIQUE
中華铁路工人
加拿大太平洋铁路公司

In the early 1880's contractor Andrew Onderdonk brought thousands of labourers from China to help build the Pacific Railway through the mountains of British Columbia. About three-quarters of the men who worked on the section between the Pacific and Craigellachie were Chinese. Although considered excellent workers, they received only a dollar a day, half the pay of a white worker. Hundreds of Chinese died from accidents or illness, for the work was dangerous and living conditions poor. Those who remained in Canada when the railway was completed securely established the basis of British Columbia's Chinese community.

Vers 1880, l'entrepreneur Andrew Onderdonk fit venir des milliers de Chinois pour construire le tronçon du chemin de fer du Pacifique à travers les montagnes de la Colombie-Britannique jusqu'à Craigellachie. Bien que ces Chinois, qui constituèrent 75% de la main-d'oeuvre, fussent d'excellents travailleurs, on ne leur versait qu'un dollar par jour, soit la moitié du salaire d'un travailleur blanc. Des centaines moururent à la suite de maladies ou d'accidents de travail. A la fin du projet, beaucoup rentrèrent en Chine. Ceux qui restèrent constituèrent les racines de l'actuelle communauté chinoise de la Colombie-Britannique.

[illegible]

Historic Sites and Monuments Board of Canada.
Commission des lieux et monuments historiques du Canada.

Government of Canada - Gouvernement du Canada

K Sasquatch Incidents Map — British Columbia

Sasquatch sighting and track reports plotted by John Green up to about 1980. Although not current, the map provides a valid appreciation of the distribution of reports throughout the province.

Bibliography

Books

Barlee, N.L., 1978. *The Best of Canada West.* Stagecoach Publishing Co. Ltd., Langley, BC, p. 58.

Green, John, 2006. *Sasquatch the Apes Among* Us. Hancock House Publishers Ltd., Surrey, BC, pp. 85–88.

Hunter, Don with René Dahinden, 1993. *Sasquatch: The Search for North America's Incredible Creature.* McClelland & Stewart, Toronto, ON, (originally published 1973), p. 23.

Krantz, Grover, Dr., 1999. *Bigfoot/Sasquatch Evidence.* Hancock House Publishers, Surrey, BC (originally published 1992, Johnson Books, Boulder, CO), pp. 202–204.

Laforet, Andrea and Annie York, 1998. *Spuzzum: Fraser Canyon Histories, 1808–1939.* UBC Press, Vancouver, BC, pp. 87, 88.

Lindsay, F.W., *Cariboo Yarns* (no date or publisher shown), p. 14.

Murphy, Christopher L., 2010. *Know the Sasquatch, Sequel & Update to Meet the Sasquatch.* Hancock House Publishers Ltd., Surrey, BC, pp. 28, 29.

Yale & District Historical Society. 2007. *Historic Yale, British Columbia.*

Newspaper Articles

"What is It? A Strange Creature Captured Above Yale." *Daily Colonist,* Victoria, BC, Canada, July 4, 1884.

"Chinese Petition." *Daily Colonist,* Victoria, BC, Canada, July 4, 1884.

"The What Is It." *Mainland Guardian,* New Westminster, BC, Canada, July 9, 1884.

"The Wild Man." *The Columbian,* New Westminster, BC, Canada, July 12, 1884.

Photographs & Images — Credits/Copyrights

Page	Description	Photo Credit/ Copyright
Front cover	Jacko in boxcar	Duncan Hopkins
2	Railway tunnel	B. Blount
6	T. Steenburg	C. Murphy
8	Jacko on tracks	Duncan Hopkins
8	Jacko in boxcar	Duncan Hopkins
10	Chief Khahtsahlano	Public Domain
11	Steamboat	Public Domain
11	Morse Lister	C. Murphy
12.	Docking facility	C. Murphy
12	B. Blount with ring	C. Murphy
13	Trail	C. Murphy
13	Rotting wood	C. Murphy
14	Cement block (top)	C. Murphy
14	Cement block (lower)	C. Murphy
15	Inside cement block	C. Murphy
16	C. Murphy and Jail	C. Murphy
16	Bunks in jail	C. Murphy
18	John Green	C. Murphy
18	René Dahinden	C. Murphy
19	Railway tunnel	B. Blount
19	Railway view	C. Murphy
22	Joe Haslam	C. Murphy
23	Railway dock	C. Murphy
23	MacInnes house	C. Murphy
23	Murphy & Chief Bobb	C. Murphy
24	Machine shops	C. Murphy
24	Machine shop location	C. Murphy

Page	Description	Photo Credit/ Copyright
25	Chinese workers	Public Domain
31	*Daily Colonist* page	Public Domain
34	*Vancouver Sun* article	*Vancouver Sun*
35	Tombstone	C. Murphy
35	Midwives list	Public Domain
36	Yale street scene	Public Domain
36	Men playing cards	Public Domain
37	Church (top)	Public Domain
37	Church (lower)	C. Murphy
38	Map	Google Earth (as shown on image)
39	Railway tracks	C. Murphy
39	House	C. Murphy
39	Old general store	C. Murphy
40	Petition – *Daily Colonist*	Public Domain
41	Monument to Chinese	C. Murphy
41	Monument plaque	C. Murphy
42	Sighting map	J. Green
Back cover (top)	C. Murphy	Marquie Murphy
Back cover (lower)	B. Blount	C. Murphy

General Index

Other **Hancock House** cryptozoology titles

Best of Sasquatch Bigfoot
John Green
0-88839-546-9
8½ x 11, sc, 144 pages

Bigfoot Discovery Coloring Book
Michael Rugg
0-88839-592-2
8½ x 11, sc, 24pages

Bigfoot Encounters in Ohio
C. Murphy, J. Cook, G. Clappison
0-88839-607-4
5½ x 8½, sc, 152 pages

Bigfoot Encounters in New York & New England
Robert Bartholomew
Paul Bartholomew
978-0-88839-652-5
5½ x 8½, sc, 176 pages

Bigfoot Film Controversy
Roger Patterson, Christopher Murphy
0-88839-581-7
5½ x 8½, sc, 240 pages

Bigfoot Film Journal
Christopher Murphy
0-88839-658-7
5½ x 8½, sc, 106 pages

Bigfoot Research: The Russian Vision
Dmitri Bayanov
978-0-88839-706-5
5½ x 8½, sc, 432 pages

Bigfoot Sasquatch Evidence
Dr. Grover S. Krantz
0-88839-447-0
5½ x 8½, sc, 348 pages

Giants, Cannibals & Monsters
Kathy Moskowitz Strain
0-88839-650-3
8½ x 11, sc, 288 pages

Hoopa Project
David Paulides
0-88839-653-2
5½ x 8½, sc, 336 pages

In Search of Giants
Thomas Steenburg
0-88839-446-2
5½ x 8½, sc, 256 pages

In Search of Ogopogo
Arlene Gaal
0-88839-482-9
5½ x 8½, sc, 208 pages

Know the Sasquatch
Christopher Murphy
978-0-88839-657-0
8½ x 11, sc, 320 pages

The Locals
Thom Powell
0-88839-552-3
5½ x 8½, sc, 272 pages

Meet the Sasquatch
Christopher Murphy, John Green, Thomas Steenburg
0-88839-574-4
8½ x 11, hc, 240 pages

Raincoast Sasquatch
J. Robert Alley
978-0-88839-508-5
5½ x 8½, sc, 360 pages

Rumours of Existence
Matthew A. Bille
0-88839-335-0
5½ x 8½, sc, 192 pages

Sasquatch: The Apes Among Us
John Green
0-88839-123-4
5½ x 8½, sc, 492 pages

Sasquatch Bigfoot
Thomas Steenburg
0-88839-312-1
5½ x 8½, sc, 128 pages

Sasquatch/Bigfoot and the Mystery of the Wild Man
Jean-Paul Debenat
978-0-88839-685-3
5½ x 8½, sc, 428 pages

Shadows of Existence
Matthew A. Bille
0-88839-612-0
5½ x 8½, sc, 320 pages

Strange Northwest
Chris Bader
0-88839-359-8
5½ x 8½, sc, 144 pages

Tribal Bigfoot
David Paulides
978-0-88839-687-7
5½ x 8½, sc, 336 pages

UFO Defense Tactics
A.K. Johnstone
0-88839-501-9
5½ x 8½, sc, 152 pages

Who's Watching You?
Linda Coil Suchy
0-88839-664-8
5½ x 8½, sc, 408 pages

View all **Hancock House** *titles at* **www.hancockhouse.com**